S0-DQW-545

Published by Creative Education
123 South Broad Street, Mankato, Minnesota 56001
Creative Education is an imprint of The Creative Company

Designed by Stephanie Blumenthal

Photographs by American Peanut Council, Archive Photos, Lisa Christenson, Cheryl Ertelt, Georgia Peanut Council, Tom Myers, Bonnie Sue Rauch, Karlene Schwartz, Tom Stack & Associates (Inga Spence), The Thacker Group, Unicorn Stock Photos (Martin Jones, Fred Reischl, Jim Shippee)

Library of Congress Cataloging-in-Publication Data

Halfmann, Janet.
Peanuts / by Janet Halfmann.
p. cm. — (Let's investigate)
Includes index.
ISBN 1-58341-191-7
1. Peanuts—Juvenile literature. [1. Peanuts.] I. Title. II. Let's investigate (Mankato, Minn.)
SB351.P3 H36 2001
633.3'68—dc21 00-047388

First edition

2 4 6 8 9 7 5 3 1

PEANUTS

JANET HALFMANN

Creative Education

PEANUT

PROTEIN

Because 26 percent of a peanut is ***protein****, the U.S. Food Guide Pyramid puts it in the Meat and Alternatives food group, along with meat, fish, poultry, and beans.*

From their original home in South America, peanuts have traveled across the ocean and back again to become one of the world's most popular foods. They're munched as snacks and stirred into main dishes and sweets. Peanut butter is a favorite of young and old alike. Cooks prize peanut oil around the globe. Not only are peanuts tasty, but they're powerhouses of nutrition.

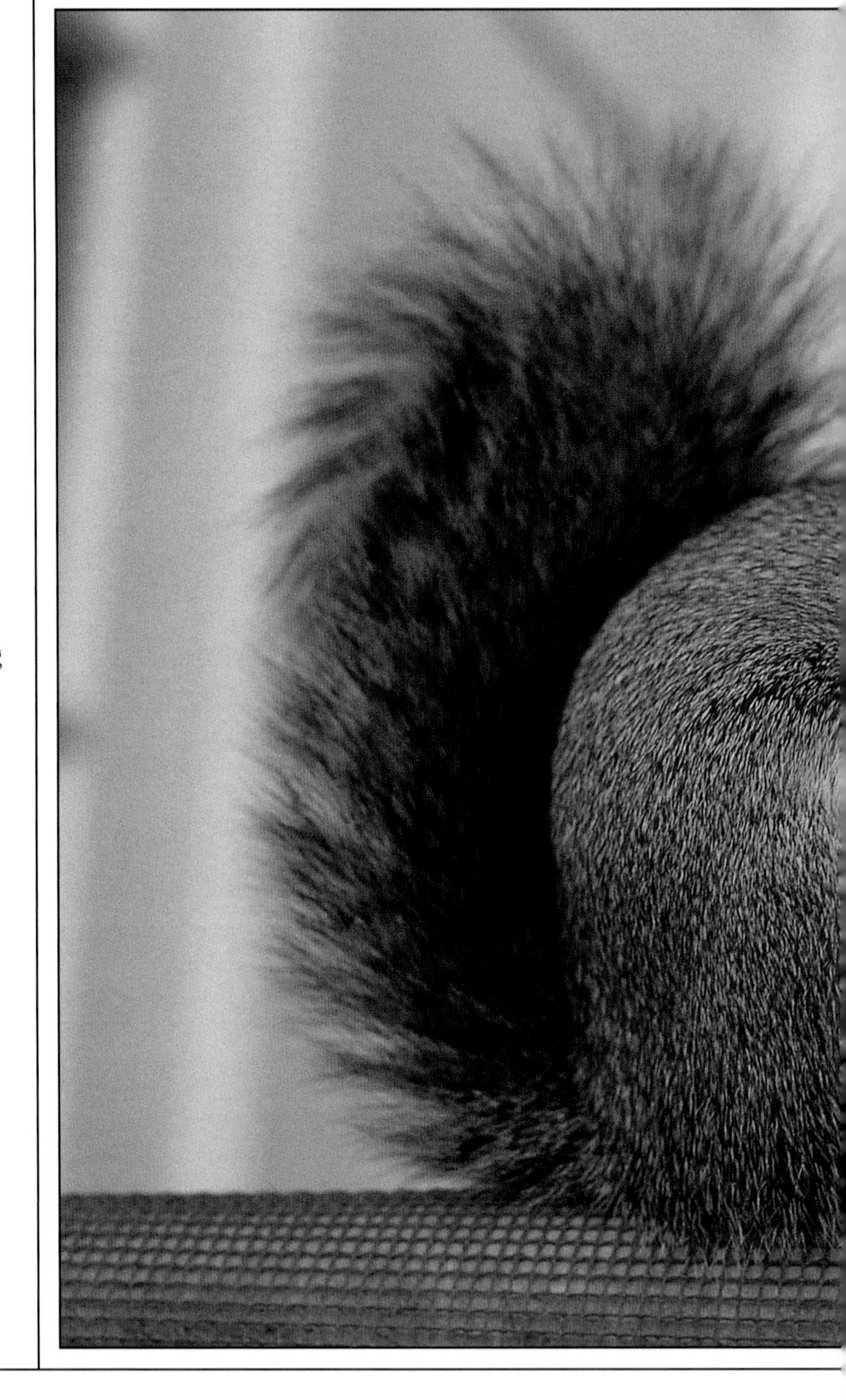

Right, a squirrel eating a peanut
Below, harvested nuts and legumes

PEANUT LAW

In Massachusetts, an old law states that it is illegal to eat peanuts in church.

PEANUT BASEBALL

A stadium owner once threatened to ban peanuts at baseball games because sweeping up the shells cost too much. Fans got upset and protested, so the owner gave away free peanuts at the following season's home opener.

PEANUT STATES

Most U.S. peanuts are grown in nine southern states: Alabama, Florida, Georgia (which grows the most), New Mexico, North Carolina, Oklahoma, South Carolina, Texas, and Virginia.

Above, former U.S. president Jimmy Carter was a peanut farmer Right, a peanut field at harvest time

THE WORLD OF PEANUTS

Most peanuts grow in the warm areas of Asia, Africa, North and South America, and Australia. Major peanut-growing countries include China, India, the United States, Nigeria, Senegal, Argentina, and Indonesia. Every year, farmers in more than 100 countries harvest about 32 million tons (29 million t) total of peanut pods.

Peanuts are high in protein and energy. Pound for pound, they contain more protein than meat and more food energy than sugar. Peanuts also have 13 vitamins and minerals.

Millions of people call the peanut their favorite nut, but the peanut isn't a nut. It's part of the legume family, like peas and beans. Legumes have **pods**, or shells containing seeds.

But the peanut differs from its pea and bean cousins in a major way—its pods grow underground. For that reason, the peanut is called a groundnut in many parts of the world. Growers around the globe often call true nuts, such as walnuts, "tree nuts" to differentiate them from peanuts.

PEANUT FACT

Legumes are second only to grains, such as rice and wheat, as the world's most popular source of food for humans.

PEANUT CAPITAL

Dotham, Alabama, claims to be the peanut capital of the world. It is the home of the yearly National Peanut Festival.

Two peanuts inside their pod

PEANUT VARIETIES

Ecuador, where peanuts have grown for centuries, may have more varieties than any other country. Scientists study these native plants for genes that may improve the varieties grown by farmers around the world.

Peanuts need a long, warm growing season and moderate rain. They grow best in loose, sandy soil. The seeds grow into plants about 18 inches (46 cm) tall with fuzzy stems and small oval leaves. The plants' golden-yellow flowers look like tiny butterflies and bloom near the bottom of the stems. The flowers **pollinate** themselves.

Peanut blossoms

As the flowers wilt, their bases grow toward the ground. No other plant does this. Each downward-growing stalk is called a peg. The peg's pointed tip pokes into the ground, turns sideways, and grows into a peanut pod. Each plant produces 40 or more pods, with two to six seeds, or peanuts, in each shell.

PEANUT NAME

The peanut's scientific name, Arachis hypogaea, *is Greek for "burying one's head." In China, the peanut is called* lo hua sheng, *which means "dropping flower gives birth."*

Above, rows of peanut plants

PEANUT CASH

Groundnuts are the number one cash crop in Gambia, West Africa. Farmers in Senegal and many other African countries also depend on groundnuts as an important cash crop, as do countries elsewhere in the world.

Many farmers harvest peanuts with machines. First, the plants are pulled from the ground and turned upside down so the sun can dry the pods. A few days later, a large harvesting machine called a combine pulls the pods from the plants and dumps them into special drying wagons. Some farmers and gardeners do these same operations by hand.

Top, inverting peanut plants for harvest Bottom, dumping pods into a drying wagon

PEANUT

SEASONINGS

To salt peanuts in the shell, salted water is forced through them. When the peanuts are roasted, the water evaporates, but the salt remains inside the shells.

The part of the peanut used for food is called the kernel. Different plants produce different kinds of kernels, or seeds. Spanish peanuts are small with reddish-brown skins. They are often used in candy. Runners are medium-sized and often used to make peanut butter. Virginias are the largest peanuts. They are often roasted in the shell, or shelled and salted. Many snack mixes include Virginias. The sweetest peanuts are Valencias. Each shell has three or more small kernels with bright red skins. They are often roasted in the shell or blanched.

Freshly salted Spanish peanuts

PEANUT SNACK

Scientists found handfuls of peanuts in women's sewing baskets in a burial site of the Ancón people, who lived from 750 to 500 B.C. along the central coast of what is now Peru. The discovery indicated that the women probably snacked on peanuts while sewing.

PEANUT HISTORY

Peanuts are one of the world's oldest **crops**. The first ones were grown about 5,000 years ago in South America. Ancient peoples of Peru buried jars of peanuts with their dead. Scientists have found peanuts in countless graves. The Moche people, who lived along Peru's northern coast from 200 B.C. to 800 A.D., included pottery with peanut designs in their burial sites. The care with which they crafted the designs shows their high esteem for the peanut.

Half-grown peanut plants

By the time Europeans explored the **New World**, Native Americans were growing peanuts throughout South America, the Caribbean, and Mexico. Peanuts were a common food, and people prepared them in a variety of ways: boiled, roasted, and broiled. They were also crushed to eat as porridge, to mix with other foods, or to make into cakes. The explorers liked roasted peanuts as snacks, so they took plants back to Europe.

PEANUT MEDICINE

Centuries ago, in what is now central Mexico, the Aztecs soothed painful, swollen gums with peanut paste.

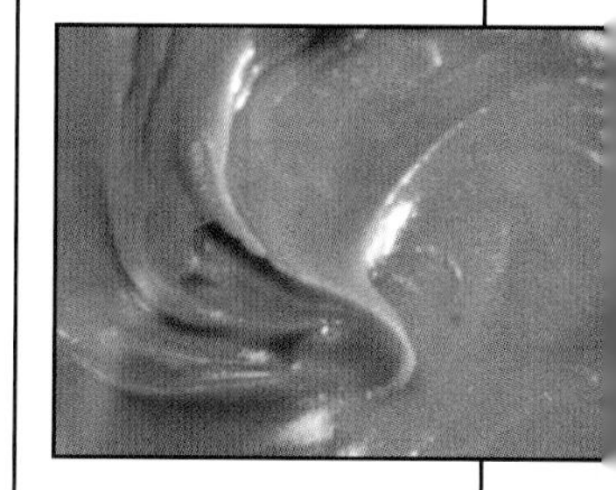

Above, early peanut paste resembled today's peanut butter

PEANUT

FOLKTALES

Peanuts, or goobers, are often a featured food in African-American folktales. For example, Brer Rabbit's fondness for goobers gets him into lots of trouble.

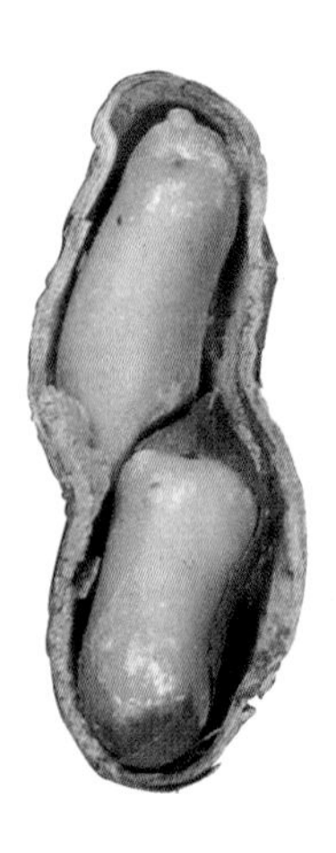

Above, salted peanuts in their shell
Right, bags of peanuts ready for shipment

Later, European traders and missionaries took peanuts to Africa and Asia, where they soon became popular. They were welcomed as a valuable source of protein and found their way into everyday cooking. For a long time, scientists believed peanuts originated in Africa and Asia because of the large quantities grown there.

Peanuts came back to the New World with the African slave trade. Slaves planted peanuts next to their cabins throughout the southern United States. They called them goober peas, or goobers, after the Kimbundu word *nguba*, the name for groundnut in the African country of Angola. Peanuts are still sometimes called goobers. The peanut also has had many other names, including ground pea and earth nut.

PEANUT COUNTING

An 18-ounce (510 g) jar of peanut butter contains an average of 850 peanuts.

Peanuts are also called goobers

PEANUT

SOIL

*How do peanuts and other legumes rebuild the soil? Special **bacteria** live in tiny growths on their roots. These bacteria take **nitrogen** out of the air and add it to the soil.*

Above, peanut pods still attached to their stems and roots
Right, a peanut field

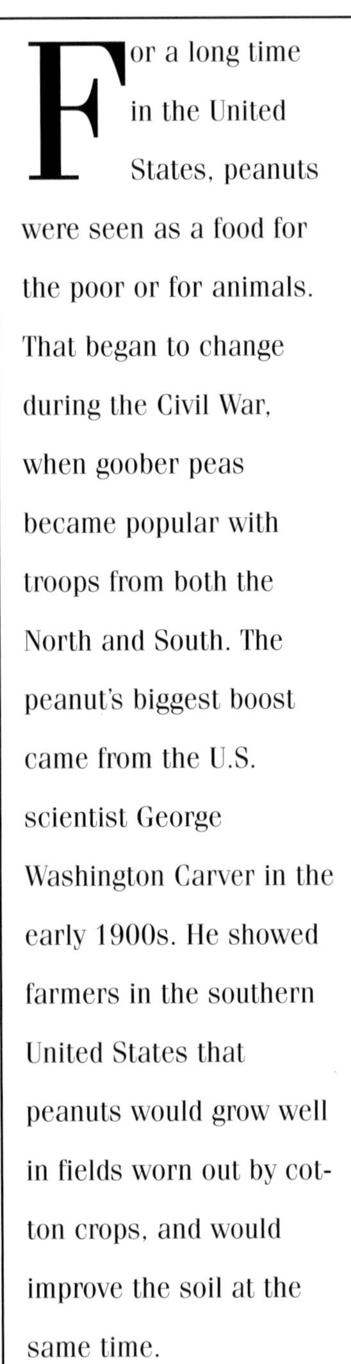

For a long time in the United States, peanuts were seen as a food for the poor or for animals. That began to change during the Civil War, when goober peas became popular with troops from both the North and South. The peanut's biggest boost came from the U.S. scientist George Washington Carver in the early 1900s. He showed farmers in the southern United States that peanuts would grow well in fields worn out by cotton crops, and would improve the soil at the same time.

PEANUT
MEAL

George Washington Carver and his students created a nine-dish feast to convince business guests of the value of peanuts. They served soup, salad, imitation chicken, a vegetable, bread, ice cream, candy, cookies, and coffee—all containing peanuts.

Carver also developed more than 300 uses for peanuts. They could be made into foods, soap, plastic, and cosmetics. Carver's work meant that more farmers could sell this new crop and make a good living. Today, George Washington Carver is known as the father of the U.S. peanut industry, and his ideas continue to be used around the world.

George Washington Carver at work

Peanuts arrived in Australia in the 1870s with the gold rush. Chinese gold prospectors planted the first peanut crops in northern Queensland. But peanut farming didn't take off in Australia until the 1920s, when a margarine company provided a market for the crop.

PEANUT
GROWERS

China and India together grow more than half of the world's peanuts.

PEANUT
GALLERY

The cheap seats high up in a theater were originally called the "peanut gallery." There, the audience munched on peanuts and tossed them at the actors if they didn't like what was happening onstage.

Overturned peanut plants drying in the sun

PEANUT

MEAL

A peanut butter sandwich, a glass of milk, and an orange provide all the nutrition for a balanced meal.

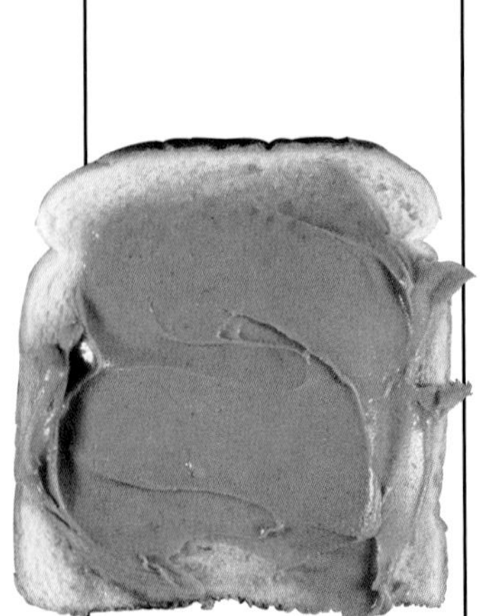

Above, bread with peanut butter
Right, healthy snacks, including a jar of home-made peanut butter

PEANUT BUTTER

Every second, someone in the United States or Canada buys a jar of peanut butter. In a year, that's enough peanut butter to cover the floor of the Grand Canyon. To help keep all those peanut butter lovers happy, the United States uses half of its peanut crop every year to make peanut butter. Canada, Australia, the Philippines, and other countries also make peanut butter. Peanut butter lovers can be found around the world. One international peanut butter fan club has 60,000 members.

Ancient South American Indians were the first to grind peanuts into a paste resembling peanut butter. They mixed it with cocoa and honey. Africans have been making peanut paste for centuries. A U.S. doctor created the modern version of peanut butter in 1890 for his elderly patients. He wanted a **nutritious** food that was easy to chew. Peanut butter was introduced to the public at the 1904 World's Fair.

PEANUT HAMS

Queen Victoria of England had a standing order that six Virginia hams from America be used every week in the palace. She liked the taste of the hams because they came from pigs that had been fed peanuts.

PEANUT BUTTER

Peanut butter lovers have been known to put peanut butter on corn on the cob, orange slices, chili, and hot apple pie.

Unshelled peanuts

PEANUT

RECIPE

To make homemade old-fashioned peanut butter, blend one cup (240 ml) of shelled, roasted, salted peanuts and one and a half teaspoons (7 ml) of peanut oil in a blender or food processor. Although additional peanut oil isn't used at the factory, it makes blending easier at home.

In a peanut butter factory, peanuts are roasted and stripped of their skins. Then, they are ground and mixed with a sweetener, salt, and a special vegetable oil that hardens to keep the peanut oil from separating. Crunchy peanut butter is made by adding chopped peanuts. Peanut butter with no added vegetable oil, and often no sugar or salt, is called old-fashioned or natural peanut butter. Its peanut oil rises to the top and needs to be mixed in before it is used.

Filling jars at a peanut butter factory

LOTS OF PEANUT OIL

Many Native Americans and Africans pressed oil from peanuts from earliest times, but the first mill to extract peanut oil wasn't built until around 1800 in Spain. Mills in England and France followed, with most of the peanuts coming from Africa. Today, more than half of all peanuts grown world-wide are pressed into oil.

PEANUT OIL

Peanut oil from West Africa was sent to France during the winter of 1830 when the country's olive trees were damaged.

Each peanut seed is about half oil, and peanut oil is the fourth most important vegetable oil in the world. It is highly prized for cooking, especially by the Chinese. They like its unique flavor and its ability to get very hot without smoking.

Peanut oil is an ingredient in many salad oils, dressings, margarines, and shortenings. In Indonesia and India, peanut oil is burned in lamps.

PEANUT LAW

U.S. law requires that peanut butter contain at least 90 percent peanuts. A product with a lower percentage must use a different name, such as peanut spread.

Above, circus elephants are often fed peanuts Left, harvesting peanuts by hand in China

PEANUT

SOIL

Peanuts are used today in Vietnam and other countries to improve the soil. In the West African country of Ghana, farmers plant peanuts between rows of yams to add nitrogen to the soil.

THE MAIN COURSE

In Africa, groundnuts have long been an important part of everyday cooking. Many people grow them near their homes. They grind them into a paste to make stews, sauces, and soups. Wherever they traveled, Africans brought along their recipes. Today, numerous versions of groundnut stew are popular not only all across Africa, but in homes and restaurants around the world.

Peanuts make tasty sauces and soups

A garden salad garnished with peanuts

In Asia, too, groundnuts have long been used in main courses. Cooks in China, Indonesia, India, Thailand, Vietnam, and other countries crush them into creamy sauces, toss them into stir-fries and salads, or add them to curries. Today, many peanut-flavored Asian dishes are popular around the globe.

In many countries, both children and adults drink peanut milk, which is high in protein. Peanut flour is added to baby foods, breads, and high-protein bars to pack them with nutrition. Scientists around the world are researching other foods to pack with peanut protein, including hot dogs, Chinese noodles, and cheese-flavored peanut spreads.

PEANUT DRINK

A rich drink made of homogenized peanut oil once was used in hospitals to strengthen patients before or after an operation. Each sweet glassful contained about 1,000 calories.

PEANUT SANDWICH

The world's largest peanut butter and jelly sandwich was 40 feet (12 m) long and spread with 150 pounds (68 kg) of peanut butter and 50 pounds (23 kg) of jelly. Its home? A Pennsylvania town called Peanut.

Above, a crunchy snack called peanut brittle Right, peanut M&M's are a top-selling candy

SNACKS AND SWEETS

Eating peanuts as snacks began centuries ago and is still going strong. Peanuts are the perfect snack—cheap, tasty, easy to carry, and nutritious. Most peanuts are roasted to bring out their flavor. Some are roasted in the shell, but most are shelled and then roasted and seasoned. Peanuts can be roasted in hot oil or dry-roasted with hot forced air. Fresh peanuts that are boiled in saltwater are called boiled peanuts.

For centuries, peanuts have been used in sweets. In West Africa, a popular sweet treat is kanya, a bar made of peanut butter, sugar, and rice cereal. The Spaniards have enjoyed chocolate-covered peanuts since the early 1800s. Six of the top 10 candy bars in the United States contain peanuts. The Chinese commonly use ground peanuts mixed with sugar as a filling in sweets.

PEANUT SNACK

In Laos, whole peanuts are stuffed into the bodies of headless grasshoppers and then roasted and sold on the street as snacks.

Peanut butter cookies shaped like peanuts

PEANUT FUTURE

Scientists are studying whether astronauts could grow peanuts for food during long space trips. They also are testing whether peanuts will grow in the soils of other planets.

MORE THAN FOOD

The non-food uses for peanuts are almost endless: shaving creams, shampoos, lipsticks, paints, lubricating oils, furniture polishes, insecticides, adhesives, dyes, explosives, inks, rubber, and medicines, to name a few. No part of the peanut plant is wasted. Peanut shells are used to make wallboard, fireplace logs, and kitty litter. Peanut skins are turned into paper. The dried vines are fed to animals or used as fertilizer, as are the peanut cakes left over from oil processing.

Paint made from peanut oil

The peanut is likely to become even more popular as scientists find still more uses for it in food and other products. But its popularity is already assured. All around the world, the peanut has become an important part of meals, fun, and everyday life.

Parade costumes covered with unshelled peanuts

PEANUT ALLERGY

Some people have a ***food allergy*** *to peanuts. Just a taste can cause a serious reaction, such as difficulty breathing or even death.*

PEANUT BITES

Dry, bite-sized peanut butter sandwiches went into space with the Apollo astronauts in the late 1960s. The astronauts' saliva added moisture to the tiny sandwiches.

Glossary

Bacteria are tiny living cells that can be seen only with the aid of a microscope.

Plants grown for food or to be sold for profit are called **crops**.

A **food allergy** is a reaction to an otherwise harmless food, usually a protein. The body perceives the food as an invader and releases chemicals to fight it. Symptoms can include hives, swelling of the throat, and constriction in the lungs, and may even result in death.

The **New World** is made up of the land in North, Central, and South America.

All plants need **nitrogen** to grow. Most plants take nitrogen from the soil, but peanuts and other legumes add it to the soil.

Food that is **nutritious** provides many of the elements needed for life and growth—for example, vitamins and minerals.

Pods are the fruits of the peanut plant. They contain the seeds, which are called peanuts.

To **pollinate** is to transfer flower dust, called pollen, from the male part of a flower to the female part so that the plant can produce seeds.

Protein is a substance that people and animals need in order to stay alive. Peanuts are high in protein.

Index